Yeas

Yeast

by Thomas Henry Huxley

I HAVE selected to-night the particular subject of Yeast for two reasons—or, rather, I should say for three. In the first place, because it is one of the simplest and the most familiar objects with which we are acquainted. In the second place, because the facts and phenomena which I have to describe are so simple that it is possible to put them before you without the help of any of those pictures or diagrams which are needed when matters are more complicated, and which, if I had to refer to them here, would involve the necessity of my turning away from you now and then, and thereby increasing very largely my difficulty (already sufficiently great) in making myself heard. And thirdly, I have chosen this subject because I know of no familiar substance forming part of our every-day knowledge and experience, the examination of which, with a little care, tends to open up such very considerable issues as does this substance—yeast.

In the first place, I should like to call your attention to a fact with which the whole of you are, to begin with, perfectly acquainted, I mean the fact that any liquid containing sugar, any liquid which is formed by pressing out the succulent parts of the fruits of plants, or a mixture of honey and water, if left to itself for a short time, begins to undergo a peculiar change. No matter how clear it might be at starting, yet after a few hours, or at most a few days, if the temperature is high, this liquid begins to be turbid, and by-and-by bubbles make their appearance in it, and a sort of dirty-looking yellowish foam or scum collects at the surface; while at the same time, by degrees, a similar kind of matter, which we call the "lees," sinks to the bottom.

The quantity of this dirty-looking stuff, that we call the scum and the lees, goes on increasing until it reaches a certain amount, and then it stops; and by the time it stops, you find the liquid in which this matter has been formed has become altered in its quality. To begin with it was a mere sweetish substance, having the flavour of whatever might be the plant from which it was expressed, or having merely the taste and the absence of smell of a solution of sugar; but by the time that this change that I have been briefly describing to you is accomplished the liquid has become completely altered, it has acquired a peculiar smell, and, what is still more remarkable, it has gained the property of intoxicating the person who drinks it. Nothing can be more innocent than a solution of sugar; nothing can be less innocent, if taken in excess, as you all know, than those fermented matters which are produced from sugar. Well, again, if you notice that bubbling, or, as it were, seething of the liquid, which has accompanied the whole of this process, you will find that it is produced by the evolution of little bubbles of air-like substance out of the liquid; and I dare say you all know this air-like substance is not like common air; it is not a substance which a man can breathe with impunity. You often hear of accidents which take place in brewers' vats when men go in carelessly, and get suffocated there without knowing that there was anything evil awaiting them. And if you tried the experiment with this liquid I am telling of while it was fermenting, you would find that any small animal let down into the vessel would be similarly stifled; and you would discover that a light lowered down into it would go out. Well, then, lastly, if after this liquid has been thus altered you expose it to that process which is called distillation; that is to say, if you put it into a still, and collect the matters which are sent over, you obtain, when you first heat it, a clear transparent liquid, which, however, is something totally different from water; it is much lighter; it has a strong smell, and it has an acrid taste; and it possesses the same intoxicating power as the original liquid, but in a much more intense degree. If you put a light to it, it burns with a bright flame, and it is that substance which we know as spirits of wine.

Now these facts which I have just put before you—all but the last—have been known from extremely remote antiquity. It is, I hope one of the best evidences of the antiquity of the human race, that among the earliest records of all kinds of men, you find a time recorded when they got drunk. We may hope that that must have been a very late period in their history. Not only have we the record of what happened to Noah, but if we turn to the traditions of a different people, those forefathers of ours who lived in the high lands of Northern India, we find that they were not less addicted to intoxicating liquids; and I have no doubt that the knowledge of this process extends far beyond the limits of historically recorded time. And it is a very curious thing to observe that all the names we have of this process, and all that belongs to it, are names that have their roots not in our present language, but in those older languages which go back to the times at which this country was peopled. That word "fermentation" for example, which is the title we apply to the whole process, is a Latin term; and a term which is evidently based upon the fact of the effervescence of the liquid. Then the French, who are very fond of calling themselves a Latin race, have a particular word for ferment, which is 'levure'. And, in the same way, we have the word "leaven," those two words having reference to the heaving up, or to the raising of the substance which is fermented. Now those are words which we get from what I may call the Latin side of our parentage; but if we turn to the Saxon side, there are a number of names connected with this process of fermentation. For example, the Germans call fermentation—and the old Germans did so—"gahren;" and they call anything which is used as a ferment by such names, such as "gheist" and "geest," and finally in low German, "yest"; and that word you know is the word our Saxon forefathers used, and is almost the same as the word which is commonly employed in this country to denote the common ferment of which I have been speaking. So they have another name, the word "hefe," which is derived from their verb "heben," which signifies to raise up; and they have yet a third name, which is also one common in this country (I do not know whether it is common in Lancashire, but it is certainly very common in the Midland countries), the word "barm," which is derived from a root which signifies to raise or to bear up. Barm is a something borne up; and thus there is much more real relation than is commonly supposed by those who make puns, between the beer which a man takes down his throat and the bier upon which that process, if carried to excess, generally lands him, for they are both derived from the root signifying bearing up; the one thing is borne upon men's shoulders, and the other is the fermented liquid which was borne up by the fermentation taking place in itself.

Again, I spoke of the produce of fermentation as "spirit of wine." Now what a very curious phrase that is, if you come to think of it. The old alchemists talked of the finest essence of anything as if it had the same sort of relation to the thing itself as a man's spirit is supposed to have to his body; and so they spoke of this fine essence of the fermented liquid as being the spirit of the liquid. Thus came about that extraordinary ambiguity of language, in virtue of which you apply precisely the same substantive name to the soul of man and to a glass of gin! And then there is still yet one other most curious piece of nomenclature connected with this matter, and that is the word "alcohol" itself, which is now so familiar to everybody. Alcohol originally meant a very fine powder. The women of the Arabs and other Eastern people are in the habit of tinging their eyelashes with a very fine black powder which is made of antimony, and they call that "kohol;" and the "al" is simply the article put in front of it, so as to say "the kohol." And up to the 17th century in this country the word alcohol was employed to signify any very fine powder; you find it in Robert Boyle's works that he uses "alcohol" for a very fine subtle powder. But then this name of anything very fine and very subtle came to be specially connected with the fine and subtle spirit obtained from the fermentation of sugar; and I believe that the first person who fairly fixed it as the proper name of what we now commonly call spirits of wine, was the great French chemist Lavoisier, so comparatively recent is the use of the word alcohol in this specialised sense.

So much by way of general introduction to the subject on which I have to speak to-night. What I have hitherto stated is simply what we may call common knowledge, which everybody may acquaint himself with. And you know that what we call scientific knowledge is not any kind of conjuration, as people sometimes suppose, but it is simply the application of the same principles of common sense that we apply to common knowledge, carried out, if I may so speak, to knowledge which is uncommon. And all that we know now of this substance, yeast, and all the very strange issues to which that knowledge has led us, have simply come out of the inveterate habit, and a very fortunate habit for the human race it is, which scientific men have of not being content until they have routed out all the different chains and connections of apparently simple phenomena, until they have taken them to pieces and understood the conditions upon which they depend. I will try to point out to you now what has happened in consequence of endeavouring to apply this process of "analysis," as we call it, this teazing out of an apparently simple fact into all the little facts of which it is made up, to the ascertained facts relating to the barm or the yeast; secondly, what has come of the attempt to ascertain distinctly what is the nature of the products which are produced by fermentation; then what has come of the attempt to understand the relation between the yeast and the products; and lastly, what very curious side issues if I may so call them—have branched out in the course of this inquiry, which has now occupied somewhere about two centuries.

The first thing was to make out precisely and clearly what was the nature of this substance, this apparently mere scum and mud that we call yeast. And that was first commenced seriously by a wonderful old Dutchman of the name of Leeuwenhoek, who lived some two hundred years ago, and who was the first person to invent thoroughly trustworthy microscopes of high powers. Now, Leeuwenhoek went to work upon this yeast mud, and by applying to it high powers of the microscope, he discovered that it was no mere mud such as you might at first suppose, but that it was a substance made up of an enormous multitude of minute grains, each of which had just as definite a form as if it were a grain of corn, although it was vastly smaller, the largest of these not being more than the two-thousandth of an inch in diameter; while, as you know, a grain of corn is a large thing, and the very smallest of these particles were not more than the seven-thousandth of an inch in diameter. Leeuwenhoek saw that this muddy stuff was in reality a liquid, in which there were floating this immense number of definitely shaped particles, all aggregated in heaps and lumps and some of them separate. That discovery remained, so to speak, dormant for fully a century, and then the question was taken up by a French discoverer, who, paying great attention and having the advantage of better instruments than Leeuwenhoek had, watched these things and made the astounding discovery that they were bodies which were constantly being reproduced and growing; than when one of these rounded bodies was once formed and had grown to its full size, it immediately began to give off a little bud from one side, and then that bud grew out until it had attained the full size of the first, and that, in this way, the yeast particle was undergoing a process of multiplication by budding, just as effectual and just as complete as the process of multiplication of a plant by budding; and thus this Frenchman, Cagniard de la Tour, arrived at the conclusion—very creditable to his sagacity, and which has been confirmed by every observation and reasoning since—that this apparently muddy refuse was neither more nor less than a mass of plants, of minute living plants, growing and multiplying in the sugary fluid in which the yeast is formed. And from that time forth we have known this substance which forms the scum and the lees as the yeast plant; and it has received a scientific name—which I may use without thinking of it, and which I will therefore give you—namely, "Torula." Well, this was a capital discovery. The next thing to do was to make out how this torula was related to the other plants. I won't weary you with the whole course of investigation, but I may sum up its results, and they are these—that the torula is a particular kind of a fungus, a particular state rather, of a fungus or mould. There are many moulds which under certain conditions give rise to this torula condition, to a substance which is not distinguishable from yeast, and which has the same properties as yeast—that is to say, which is able to decompose sugar in the curious way that we shall consider by-and-by. So that the yeast plant is a plant belonging to a group of the Fungi, multiplying and growing and living in this very remarkable manner in the sugary fluid which is, so to speak, the nidus or home of the yeast.

That, in a few words, is, as far as investigation—by the help of one's eye and by the help of the microscope—has taken us. But now there is an observer whose methods of observation are more refined than those of men who use their eye, even though it be aided by the microscope; a man who sees indirectly further than we can see directly—that is, the chemist; and the chemist took up this question, and his discovery was not less remarkable than that of the microscopist. The chemist discovered that the yeast plant being composed of a sort of bag, like a bladder, inside which is a peculiar soft, semifluid material—the chemist found that this outer bladder has the same composition as the substance of wood, that material which is called "cellulose," and which consists of the elements carbon and hydrogen and oxygen, without any nitrogen. But then he also found (the first person to discover it was an Italian chemist, named Fabroni, in the end of the last century) that this inner matter which was contained in the bag, which constitutes the yeast plant, was a substance containing the elements carbon and hydrogen and oxygen and nitrogen; that it was what Fabroni called a vegeto-animal substance, and that it had the peculiarities of what are commonly called "animal products."

This again was an exceedingly remarkable discovery. It lay neglected for a time, until it was subsequently taken up by the great chemists of modern times, and they, with their delicate methods of analysis, have finally decided that, in all essential respects, the substance which forms the chief part of the contents of the yeast plant is identical with the material which forms the chief part of our own muscles, which forms the chief part of our own blood, which forms the chief part of the white of the egg; that, in fact, although this little organism is a plant, and nothing but a plant, yet that its active living contents contain a substance which is called "protein," which is of the same nature as the substance which forms the foundation of every animal organism whatever.

Now we come next to the question of the analysis of the products, of that which is produced during the process of fermentation. So far back as the beginning of the 16th century, in the times of transition between the old alchemy and the modern chemistry, there was a remarkable man, Von Helmont, a Dutchman, who saw the difference between the air which comes out of a vat where something is fermenting and common air. He was the man who invented the term "gas," and he called this kind of gas "gas silvestre"—so to speak gas that is wild, and lives in out of the way places—having in his mind the identity of this particular kind of air with that which is found in some caves and cellars. Then, the gradual process of investigation going on, it was discovered that this substance, then called "fixed air," was a poisonous gas, and it was finally identified with that kind of gas which is obtained by burning charcoal in the air, which is called "carbonic acid." Then the substance alcohol was subjected to examination, and it was found to be a combination of carbon, and hydrogen, and oxygen. Then the sugar which was contained in the fermenting liquid was examined and that was found to contain the three elements carbon, hydrogen, and oxygen. So that it was clear there were in sugar the fundamental elements which are contained in the carbonic acid, and in the alcohol. And then came that great chemist Lavoisier, and he examined into the subject carefully, and possessed with that brilliant thought of his which happens to be propounded exactly apropos to this matter of fermentation—that no matter is ever lost, but that matter only changes its form and changes its combinations—he endeavoured to make out what became of the sugar which was subjected to fermentation. He thought he discovered that the whole weight of the sugar was represented by the carbonic acid produced; that in other words, supposing this tumbler to represent the sugar, that the action of fermentation was as it were the splitting of it, the one half going away in the shape of carbonic acid, and the other half going away in the shape of alcohol. Subsequent inquiry, careful research with the refinements of modern chemistry, have been applied to this problem, and they have shown that Lavoisier was not quite correct; that what he says is quite true for about 95 per cent. of the sugar, but that the other 5 per cent., or nearly so, is converted into two other things; one of them, matter which is called succinic acid, and the other matter which is called glycerine, which you all know now as one of the commonest of household matters. It may be that we have not got to the end of this refined analysis yet, but at any rate, I suppose I may say—and I speak with some little hesitation for fear my friend Professor Roscoe here may pick me up for trespassing upon his province—but I believe I may say that now we can account for 99 per cent. at least of the sugar, and that 99 per cent. is split up into these four things, carbonic acid, alcohol, succinic acid, and glycerine. So that it may be that none of the sugar whatever disappears, and that only its parts, so to speak, are re-arranged, and if any of it disappears, certainly it is a very small portion.

Now these are the facts of the case. There is the fact of the growth of the yeast plant; and there is the fact of the splitting up of the sugar. What relation have these two facts to one another?

For a very long time that was a great matter of dispute. The early French observers, to do them justice, discerned the real state of the case, namely, that there was a very close connection between the actual life of the yeast plant and this operation of the splitting up of the sugar; and that one was in some way or other connected with the other. All investigation subsequently has confirmed this original idea. It has been shown that if you take any measures by which other plants of like kind to the torula would be killed, and by which the yeast plant is killed, then the yeast loses its efficiency. But a capital experiment upon this subject was made by a very distinguished man, Helmholz, who performed an experiment of this kind. He had two vessels—one of them we will suppose full of yeast, but over the bottom of it, as this might be, was tied a thin film of bladder; consequently, through that thin film of bladder all the liquid parts of the yeast would go, but the solid parts would be stopped behind; the torula would be stopped, the liquid parts of the yeast would go. And then he took another vessel containing a fermentable solution of sugar, and he put one inside the other; and in this way you see the fluid parts of the yeast were able to pass through with the utmost ease into the sugar, but the solid parts could not get through at all. And he judged thus: if the fluid parts are those which excite fermentation, then, inasmuch as these are stopped, the sugar will not ferment; and the sugar did not ferment, showing quite clearly, that an immediate contact with the solid, living torula was absolutely necessary to excite this process of splitting up of the sugar. This experiment was quite conclusive as to this particular point, and has had very great fruits in other directions.

Well, then, the yeast plant being essential to the production of fermentation, where does the yeast plant come from? Here, again, was another great problem opened up, for, as I said at starting, you have, under ordinary circumstances in warm weather, merely to expose some fluid containing a solution of sugar, or any form of syrup or vegetable juice to the air, in order, after a comparatively short time, to see all these phenomena of fermentation. Of course the first obvious suggestion is, that the torula has been generated within the fluid. In fact, it seems at first quite absurd to entertain any other conviction; but that belief would most assuredly be an erroneous one.

Towards the beginning of this century, in the vigorous times of the old French wars, there was a Monsieur Appert, who had his attention directed to the preservation of things that ordinarily perish, such as meats and vegetables, and in fact he laid the foundation of our modern method of preserving meats; and he found that if he boiled any of these substances and then tied them so as to exclude the air, that they would be preserved for any time. He tried these experiments, particularly with the must of wine and with the wort of beer; and he found that if the wort of beer had been carefully boiled and was stopped in such a way that the air could not get at it, it would never ferment. What was the reason of this? That, again, became the subject of a long string of experiments, with this ultimate result, that if you take precautions to prevent any solid matters from getting into the must of wine or the wort of beer, under these circumstances—that is to say, if the fluid has been boiled and placed in a bottle, and if you stuff the neck of the bottle full of cotton wool, which allows the air to go through and stops anything of a solid character however fine, then you may let it be for ten years and it will not ferment. But if you take that plug out and give the air free access, then, sooner or later fermentation will set up. And there is no doubt whatever that fermentation is excited only by the presence of some torula or other, and that that torula proceeds in our present experience, from pre-existing torulae. These little bodies are excessively light. You can easily imagine what must be the weight of little particles, but slightly heavier than water, and not more than the two-thousandth or perhaps seven-thousandth of an inch in diameter. They are capable of floating about and dancing like motes in the sunbeam; they are carried about by all sorts of currents of air; the great majority of them perish; but one or two, which may chance to enter into a sugary solution, immediately enter into active life, find there the conditions of their nourishment, increase and multiply, and may give rise to any quantity whatever of this substance yeast.

And, whatever may be true or not be true about this "spontaneous generation," as it is called in regard to all other kinds of living things, it is perfectly certain, as regards yeast, that it always owes its origin to this process of transportation or inoculation, if you like so to call it, from some other living yeast organism; and so far as yeast is concerned, the doctrine of spontaneous generation is absolutely out of court. And not only so, but the yeast must be alive in order to exert these peculiar properties. If it be crushed, if it be heated so far that its life is destroyed, that peculiar power of fermentation is not excited. Thus we have come to this conclusion, as the result of our inquiry, that the fermentation of sugar, the splitting of the sugar into alcohol and carbonic acid, glycerine, and succinic acid, is the result of nothing but the vital activity of this little fungus, the torula.

And now comes the further exceedingly difficult inquiry—how is it that this plant, the torula, produces this singular operation of the splitting up of the sugar? Fabroni, to whom I referred some time ago, imagined that the effervescence of fermentation was produced in just the same way as the effervescence of a sedlitz powder, that the yeast was a kind of acid, and that the sugar was a combination of carbonic acid and some base to form the alcohol, and that the yeast combined with this substance, and set free the carbonic acid; just as when you add carbonate of soda to acid you turn out the carbonic acid. But of course the discovery of Lavoisier that the carbonic acid and the alcohol taken together are very nearly equal in weight to the sugar, completely upset this hypothesis. Another view was therefore taken by the French chemist, Thenard, and it is still held by a very eminent chemist, M. Pasteur, and their view is this, that the yeast, so to speak, eats a little of the sugar, turns a little of it to its own purposes, and by so doing gives such a shape to the sugar that the rest of it breaks up into carbonic acid and alcohol.

Well, then, there is a third hypothesis, which is maintained by another very distinguished chemist, Liebig, which denies either of the other two, and which declares that the particles of the sugar are, as it were, shaken asunder by the forces at work in the yeast plant. Now I am not going to take you into these refinements of chemical theory, I cannot for a moment pretend to do so, but I may put the case before you by an analogy. Suppose you compare the sugar to a card house, and suppose you compare the yeast to a child coming near the card house, then Fabroni's hypothesis was that the child took half the cards away; Thenard's and Pasteur's hypothesis is that the child pulls out the bottom card and thus makes it tumble to pieces; and Liebig's hypothesis is that the child comes by and shakes the table and tumbles the house down. I appeal to my friend here (Professor Roscoe) whether that is not a fair statement of the case.

Having thus, as far as I can, discussed the general state of the question, it remains only that I should speak of some of those collateral results which have come in a very remarkable way out of the investigation of yeast. I told you that it was very early observed that the yeast plant consisted of a bag made up of the same material as that which composes wood, and of an interior semifluid mass which contains a substance, identical in its composition, in a broad sense, with that which constitutes the flesh of animals. Subsequently, after the structure of the yeast plant had been carefully observed, it was discovered that all plants, high and low, are made up of separate bags or "cells," as they are called; these bags or cells having the composition of the pure matter of wood; having the same composition, broadly speaking, as the sac of the yeast plant, and having in their interior a more or less fluid substance containing a matter of the same nature as the protein substance of the yeast plant. And therefore this remarkable result came out—that however much a plant may differ from an animal, yet that the essential constituent of the contents of these various cells or sacs of which the plant is made up, the nitrogenous protein matter, is the same in the animal as in the plant. And not only was this gradually discovered, but it was found that these semifluid contents of the plant cell had, in many cases, a remarkable power of contractility quite like that of the substance of animals. And about 24 or 25 years ago, namely, about the year 1846, to the best of my recollection, a very eminent German botanist, Hugo Von Mohl, conferred upon this substance which is found in the interior of the plant cell, and which is identical with the matter found in the inside of the yeast cell, and which again contains an animal substance similar to that of which we ourselves are made up—he conferred upon this that title of "protoplasm," which has brought other people a great deal of trouble since! I beg particularly to say that, because I find many people suppose that I was the inventor of that term, whereas it has been in existence for at least twenty-five years. And then other observers, taking the question up, came to this astonishing conclusion (working from this basis of the yeast), that the differences between animals and plants are not so much in the fundamental substances which compose them, not in the protoplasm, but in the manner in which the cells of which their bodies are built up have become modified. There is a sense in which it is true—and the analogy was pointed out very many years ago by some French botanists and chemists—there is a sense in which it is true that every plant is substantially an enormous aggregation of bodies similar to yeast cells, each having to a certain extent its own independent life. And there is a sense in which it is also perfectly true—although it would be impossible for me to give the statement to you with proper qualifications and limitations on an occasion like this—but there is also a sense in which it is true that every animal body is made up of an aggregation of minute particles of protoplasm, comparable each of them to the individual separate yeast plant. And those who are acquainted with the history of the wonderful revolution which has been worked in our whole conception of these matters in the last thirty years, will bear me out in saying that the first germ of them, to a very great extent, was made to grow and fructify by the study of the yeast plant, which presents us with living matter in almost its simplest condition.

Then there is yet one last and most important bearing of this yeast question. There is one direction probably in which the effects of the careful study of the nature of fermentation will yield results more practically valuable to mankind than any other. Let me recall to your minds the fact which I stated at the beginning of this lecture. Suppose that I had here a solution of pure sugar with a little mineral matter in it; and suppose it were possible for me to take upon the point of a needle one single, solitary yeast cell, measuring no more perhaps than the three-thousandth of an inch in diameter—not bigger than one of those little coloured specks of matter in my own blood at this moment, the weight of which it would be difficult to express in the fraction of a grain—and put it into this solution. From that single one, if the solution were kept at a fair temperature in a warm summer's day, there would be generated, in the course of a week, enough torulae to form a scum at the top and to form lees at the bottom, and to change the perfectly tasteless and entirely harmless fluid, syrup, into a solution impregnated with the poisonous gas carbonic acid, impregnated with the poisonous substance alcohol; and that, in virtue of the changes worked upon the sugar by the vital activity of these infinitesimally small plants. Now you see that this is a case of infection. And from the time that the phenomenon of fermentation were first carefully studied, it has constantly been suggested to the minds of thoughtful physicians that there was a something astoundingly similar between this phenomena of the propagation of fermentation by infection and contagion, and the phenomena of the propagation of diseases by infection and contagion. Out of this suggestion has grown that remarkable theory of many diseases which has been called the "germ theory of disease," the idea, in fact, that we owe a great many diseases to particles having a certain life of their own, and which are capable of being transmitted from one living being to another, exactly as the yeast plant is capable of being transmitted from one tumbler of saccharine substance to another. And that is a perfectly tenable hypothesis, one which in the present state of medicine ought to be absolutely exhausted and shown not to be true, until we take to others which have less analogy in their favour. And there are some diseases most assuredly in which it turns out to be perfectly correct. There are some forms of what are called malignant carbuncle which have been shown to be actually effected by a sort of fermentation, if I may use the phrase, by a sort of disturbance and destruction of the fluids of the animal body, set up by minute organisms which are the cause of this destruction and of this disturbance; and only recently the study of the phenomena which accompany vaccination has thrown an immense light in this direction, tending to show by experiments of the same general character as that to which I referred as performed by Helmholz, that there is a most astonishing analogy between the contagion of that healing disease and the contagion of destructive diseases. For it has been made out quite clearly, by investigations carried on in France and in this country, that the only part of the vaccine matter which is contagious, which is capable of carrying on its influence in the organism of the child who is vaccinated, is the solid particles and not the fluid. By experiments of the most ingenious kind, the solid parts have been separated from the fluid parts, and it has then been discovered that you may vaccinate a child as much as you like with the fluid parts, but no effect takes place, though an excessively small portion of the solid particles, the most minute that can be separated, is amply sufficient to give rise to all the phenomena of the cow pock, by a process which we can compare to nothing but the transmission of fermentation from one vessel into another, by the transport to the one of the torula particles which exist in the other. And it has been shown to be true of some of the most destructive diseases which infect animals, such diseases as the sheep pox, such diseases as that most terrible and destructive disorder of horses, glanders, that in these, also, the active power is the living solid particle, and that the inert part is the fluid. However, do not suppose that I am pushing the analogy too far. I do not mean to say that the active, solid parts in these diseased matters are of the same nature as living yeast plants; but, so far as it goes, there is a most surprising analogy between the two; and the value of the analogy is this, that by following it out we may some time or other come to understand how these diseases are propagated, just as we understand, now, about fermentation; and that, in this way, some of the greatest scourges which afflict the human race may be, if not prevented, at least largely alleviated.

This is the conclusion of the statements which I wished to put before you. You see we have not been able to have any accessories. If you will come in such numbers to hear a lecture of this kind, all I can say is, that diagrams cannot be made big enough for you, and that it is not possible to show any experiments illustrative of a lecture on such a subject as I have to deal with. Of course my friends the chemists and physicists are very much better off, because they can not only show you experiments, but you can smell them and hear them! But in my case such aids are not attainable, and therefore I have taken a simple subject and have dealt with it in such a way that I hope you all understand it, at least so far as I have been able to put it before you in words; and having once apprehended such of the ideas and simple facts of the case as it was possible to put before you, you can see for yourselves the great and wonderful issues of such an apparently homely subject.

Yeast

Yeast

Yeast

Yeast

Yeast

Yeast

CPSIA information can be obtained at www.ICGtesting.com
Printed in the USA
LVOW10s1917111214

418274LV00052B/1016/P